# Disclaimer

The publisher of this book is by no way associated with the National Institute of Standards and Technology (NIST). The NIST did not publish this book. It was published by 50 page publications under the public domain license.

50 Page Publications.

**Book Title:** A 1D Spectral Image Validation/Verification Metric for Fingerprints

**Book Author:** John M. Libert; John D. Grantham; Shahram Orandi

**Book Abstract:** Image validation and verification are important functions in the acquisition of fingerprint images from live-scan devices and for assessing and maintaining the fidelity of fingerprint image databases. In addition to law enforcement, such databases are used by NIST and others to test automated fingerprint identification system (AFIS) algorithms and to aide the advance of this technology. Image screening by visual inspection is time consuming. We propose a computational mechanism by which to screen fingerprint image databases for specimens improperly scanned from fingerprint cards, guide the auto-capture process and flag auto-capture failures, identify non-fingerprint images that may have been included in a database, and recognize aberrant sampling of fingerprint images. The scheme reduces an input image to a 1-dimensional power spectrum that makes explicit the characteristic ridge structure of the fingerprint that on a global basis differentiates it from most other images. The magnitude of the distinctive spectral feature, related directly to the distinctness of the level 1 ridge flow, provides a primary diagnostic indicator of the presence of a fingerprint image. The frequency of the spectral feature provides a secondary classification metric and, on a coarse level, indicates the scan sample rate of the fingerprint image. Test results are reported in which the Spectral Image Validation and Verification (SIVV) utility is applied to a variety of databases composed of fingerprint and non-fingerprint images. An equal error rate (EER) for false positive and false negative classifications of 10% is achieved for fingerprints mixed with a variety of non-fingerprint images and an EER of around 7% is found with a dataset containing fingerprints mixed with other biometric samples, i.e. face and iris images.

**Citation:** NIST Interagency/Internal Report (NISTIR) - 7599

**Keyword:** fingerprint image validation;biometrics;fingerprint;spectral analysis

# NISTIR 7599

# A 1D Spectral Image Validation/Verification Metric for Fingerprints

John M. Libert
John Grantham
Shahram Orandi

Information Access Division
Information Technology Laboratory

August 2009

National Institute of Standards and Technology
Technology Administration, U.S. Department of Commerce

# NISTIR 7599

# A 1D Spectral Image Validation/Verification Metric for Fingerprints

John M. Libert
John Grantham
Shahram Orandi

*U.S. Department of Commerce*
*Technology Administration*
*National Institute of Standards and Technology*
*Information Technology Lab*
*Gaithersburg, MD 20899*

# August 2009

U.S. Department of Commerce
*Gary Locke, Secretary*

National Institute of Standards and Technology
*Patrick D. Gallagher, Deputy Director*

## Acknowledgements

The authors wish to acknowledge the contributions of those who have facilitated the research and development work described in the present document. We would like to acknowledge and thank Scott Swann of the Federal Bureau of Investigation's (FBI) Criminal Justice Information Services (CJIS) as well as other key partners at the FBI who provided both sponsorship and support to NIST in facilitating this research.

## Table of Contents

ACKNOWLEDGEMENTS ..................................................................................................................... i
ABSTRACT ........................................................................................................................................... 1
1. INTRODUCTION ............................................................................................................................. 3
2. BACKGROUND ................................................................................................................................ 5
3. SIVV METHOD ................................................................................................................................. 7
    3.1    IMAGE WINDOWING ............................................................................................................ 7
    3.2    DISCRETE FOURIER TRANSFORM (DFT) ............................................................................... 9
    3.3    NORMALIZED POWER SPECTRUM ....................................................................................... 10
    3.4    POLAR TRANSFORM OF POWER SPECTRUM ......................................................................... 10
    3.5    1D NORMALIZED POWER SPECTRUM .................................................................................. 11
    3.6    PEAK STRUCTURE OF SPECTRUM ....................................................................................... 13
4. RESPONSE OF THE SIGNAL TO IMAGE PARAMETERS ..................................................... 17
    4.1    ROTATION INVARIANCE ..................................................................................................... 17
    4.2    RESPONSE OF SPECTRUM TO TRANSLATION ....................................................................... 17
    4.3    SCAN SAMPLE RATE SENSITIVITY (1000 PPI VS. 500 PPI) .................................................. 20
5. RESULTS AND DISCUSSION ...................................................................................................... 23
6. CONCLUSIONS ............................................................................................................................... 29
7. REFERENCES ................................................................................................................................. 31
8. APPENDIX A: DEMONSTRATION OF ROTATION INVARIANCE .................................... 33
9. APPENDIX B: TEST DATA .......................................................................................................... 35
10. APPENDIX C: DISTRIBUTIONS OF SIVV FEATURES ....................................................... 37

## List of Figures

Fig. 1 2D Blackman window function..................................................................................8

Fig. 2 Input image (995 x 1090 pixels) and Blackman windowed version.......................8

Fig. 3 Sample fingerprint image (922 x 860 pixels) and associated 1D spectra computed without and with application of a window function....................................................9

Fig. 4 Diagram showing operation performed by a function that shifts quadrants of the DFT such that dc term, $P(0,0)$, is centered in the image..............................10

Fig. 5 Illustration of processing steps from image to 1D Normalized Power Spectrum. (Image of size 995 x 1090 pixel) Spatial frequency, $u$ and $v$, are in units of cycles per image width as are the unnormalized units of polar radius, $\rho$. Polar angles, $\theta$, are in degrees. In the log power spectrum, radius values of cycles/image width are scaled to become cycles/pixel..................................................12

Fig. 6 Fingerprint sampled at 1000 ppi (1107 x 1145 pixels) and associated 1D spectrum. The pair of local minimum and maximum having the largest difference in power are indicated with green and red marks...................................14

Fig. 7 Fingerprint sampled at 1000 ppi (1990 x 790 pixels) and associated 1D spectrum..............................................................................................................14

Fig. 8 Slap-four image sampled at 1000 ppi (2049 x 3116) and spectrum......................15

Fig. 9 Off-center fingerprint image (1500 x 1600) and associated spectra computed both without and with Blackman windowing..........................................................18

Fig. 10 Image from fig 9 with fingerprint moved toward more centered position applying a circular shift to the image................................................................18

Fig. 11 Fingerprint image digitized at 1000 ppi (995 x 1090 pixels)................................21

Fig. 12 Fingerprint image from fig. 11 resampled at 500 ppi (498 x 545 pixels) using bicubic interpolation. The spectral features are shifted by factor of 2 in frequency...........................................................................................................21

Fig. 13 Box plot showing distributions of peak height for non-fingerprint images (0) and fingerprint images (1) in a mixed image dataset..............................................................................................................24

Fig. 14 Box plot showing distributions of frequency location of the valley-peak structure for non-fingerprint images (0) and fingerprint images (1) in a mixed image dataset..............................................................................................................24

Fig. 15 Plot of false positive vs. false negative error rates with variation of threshold value of peak height used as a classifier applied to a mixed image dataset. Feature frequency maximum was set at 0.15 cycles/pixel...........................................................................................................25

Fig. 16 Box plot showing distributions of peak height ($dy$) for equal numbers of fingerprint, face, and iris images in a mixed biometric dataset............................26

Fig. 17 Box plot showing distributions of frequency location of the valley-peak feature for equal numbers of fingerprint, face, and iris images in a mixed biometric dataset..................................................................................................................26

Fig. 18 Peak height ($dy$) vs. frequency of SIVV output for images of the Mixed Biometric 1 dataset. Note the separation of fingerprints into clusters corresponding to the 1000 and 500 ppi samples..........................................................27

Fig. 19 Plot of false positive vs. false negative error rates with variation of threshold value of peak height used as a classifier applied to the mixed biometric dataset. Feature frequency maximum was set at 0.15 cycles/pixel.....................................27

Fig. A-1 Circular sample of a rolled fingerprint image (1029 x 1023) captured at 1000 ppi. Original orientation (top) is rotated 45° and 90°................................................33

Fig. A-2 Circular sample of a slap-four image (2012 x 2064) captured at 1000 ppi. Original orientation (top) is rotated 45° and 90°................................................34

Fig. C-1 Box plot showing distributions of peak height ($dy$) by finger for SD27 dataset sampled at 1000 ppi. Fingers 1 – 10 are rolled inked prints on FBI 10-print cards; 11 and 12 are plain impressions of thumbs; and 13 and 14 are slap-four prints of left and right hands.................................................................................38

Fig. C-2 Box plot showing distributions of peak height ($dy$) by finger for SD27 dataset down-sampled (with anti-aliasing) to 500 ppi. Fingers 1 – 10 are rolled inked prints on FBI 10-print cards; 11 and 12 are plain impressions of thumbs; and 13 and 14 are slap-four prints of left and right hands.................................................38

Fig. C-3 Box plot showing distributions of peak height ($dy$) by finger for SD29 dataset sampled at 500 ppi. Fingers 1 – 10 are plain inked prints.....................................39

Fig. C-4 Box plot showing distributions of peak frequency location of the valley-peak feature by finger for SD27 dataset sampled at 1000 ppi. Fingers 1 – 10 are rolled inked prints on FBI 10-print cards; 11 and 12 are plain impressions of thumbs; and 13 and 14 are slap-four prints of left and right hands.....................................39

Fig. C-5 Box plot showing distributions of frequency location of the valley-peak feature by finger for SD27 dataset down-sampled (with anti-aliasing) to 500 ppi. Fingers 1 – 10 are rolled inked prints on FBI 10-print cards; 11 and 12 are plain impressions of thumbs; and 13 and 14 are slap-four prints of left and right hands......................................................................................................................40

Fig. C-6 Box plot showing distributions of frequency location of the valley-peak feature by finger for SD29 dataset sampled at 500 ppi. Fingers 1 – 10 are plain inked prints.....................................................................................................................40

## ABSTRACT

Image validation and verification are important functions in the acquisition of fingerprint images from live-scan devices and for assessing and maintaining the fidelity of fingerprint image databases. Such databases are used by law enforcement agencies, for which data integrity is paramount, and many hours must be devoted to visual inspection of images. In addition, such databases are used by the National Institute of Standards and Technology (NIST) and others to test automated fingerprint identification system (AFIS) algorithms and to aide the advance of this technology. We propose a comparatively simple computational mechanism by which to screen fingerprint image databases for specimens improperly scanned from fingerprint cards, guide the auto-capture process and flag auto-capture failures, identify non-fingerprint images that may have been included in a database, and recognize aberrant sampling of fingerprint images. The scheme reduces an input image to a 1-dimensional power spectrum that makes explicit the characteristic ridge structure of the fingerprint that on a global basis differentiates it from most other images. The magnitude of the distinctive spectral feature, related directly to the distinctness of the level 1 ridge flow, provides a primary diagnostic indicator of the presence of a fingerprint image. The frequency of the spectral feature provides a secondary classification metric and, on a coarse level, indicates the scan sample rate of the fingerprint image. Test results are reported in which the Spectral Image Validation and Verification (SIVV) utility is applied to a variety of databases composed of fingerprint and non-fingerprint images. Using only the peak height and frequency limit as simple classification criteria, the SIVV utility achieves an equal error rate (EER) for false positive and false negative classifications of 10 % for fingerprints mixed with a variety of non-fingerprint images, including many chosen to exhibit periodic structure similar to that of a fingerprint. An EER of around 7 % is found with a dataset containing fingerprints mixed with other biometric samples, i.e. face and iris images.

# 1. INTRODUCTION

The present investigation explores the use of a 1D summary of the 2D image spectrum to test validity of fingerprint images and to diagnose acquisition problems such as aberrant sampling, such as misplaced fingers in a single finger capture, or faulty segmentation. The preparation of datasets of biometric samples for various development and testing applications as well as for law enforcement and homeland security applications requires that the images be screened with respect to their validity as examples of the expected biometric. Starting with fingerprint acquisition using a live-scan device, detection of the presence of a finger or the position of the finger on the platen should aid in automating such a system. A rapid fingerprint recognition capability could serve as a preprocess for a more sophisticated image quality algorithm, such as the NIST Fingerprint Image Quality (NFIQ) metric [1, 2]. Such a system could assist in identifying mislabeled images in a mixed biometric dataset, for example where a face image is erroneously labeled as a fingerprint. A validation/verification signal might support identification of sampling errors, for example, in which an image fails to conform to a specified scan sample rate. Accordingly, the present research activity was undertaken to develop and test a computational scheme for fingerprint image validation and verification. Our method, computationally simple, differentiates fingerprint images on the basis of a power spectral feature related to the periodic texture of friction ridge skin. The relative height and frequency of the dominant spectral peak differentiates a fingerprint image from a variety of other image types, including other biometrics such as face and iris images. In view of its implementation and application we have named the method the Spectral Image Validation and Verification (SIVV) utility.

This report provides in Section 2 a brief survey of previous work on fingerprint image validation and verification, related application of spectral analysis, and decomposition of fingerprint images. Section 3 details the mathematical and algorithmic underpinnings of the SIVV method. Section 4 examines robustness of the diagnostic spectral pattern under rotation and translation of the fingerprint, and provides qualitative comparisons of the SIVV response to non-fingerprint images. Section 5 provides a quantitative examination of the SIVV performance as an aid in classifying fingerprints among assorted non-fingerprint images and among face and iris images. Finally, Section 6 summarizes the utility of the SIVV in differentiating fingerprint from non-fingerprint images, detecting scan sample rate anomalies, and detecting some types of live-scan acquisition errors.

## 2. BACKGROUND

Local orientation and frequency of ridges are recognized as unique intrinsic properties of fingerprint images. Some attention is directed toward examining such ridge structure in order to assess quality, enhance quality, and to evaluate suitability of fingerprint images for verification/recognition. Most of the work has involved various wavelet representations of the fingerprint image directed toward evaluation of the image quality, e.g. [3-6]. Other work such as that of Maio [7] and Yin [8] looks specifically at ridge spacing and distinctness of the pattern, mainly with respect to characterization of the fingerprint ridge structure as a distinct texture to be used in a recognition scheme as well as for evaluation of image quality.

Lim *et al* [9] describe a means by which to evaluate the quality as well as to test the validity of fingerprint images with respect to the ridge structure of fingerprint images. Local analysis of "orientation certainty" is used as an indicator of quality, while the ridge-to-valley structure provides indication of fingerprint image validity, i.e., that the image actually exhibits a fingerprint. They evaluate quality on a local level as indicated by the strength of energy concentration along ridge-valley orientations. Validity is indicated by measures of global uniformity and continuity of orientation across adjacent image blocks.

Tabassi, Wilson, and Watson [1, 2] describe a fingerprint image quality method that incorporates a measure of ridge flow based on local (blockwise) analysis of the discrete cosine transform signal. While the NIST Image Quality (NFIQ) metric itself primarily examines minutia and aims to predict the performance of fingerprint matchers, the reliability of minutia detection in each image neighborhood is assessed via local measures of ridge flow direction, contrast, and curvature.

The closest methods to those presently under examination are efforts directed toward using the spectral characteristics of images for texture-based image retrieval. Using a mechanism somewhat similar to that proposed in the present paper is the RAH (Radius Angle Histogram) described by Wang *et al* [10]. These researchers describe use of power spectral histograms for texture-based image retrieval. They construct the polar representation of the 2D power spectrum and characterize image textures via histograms of the quantized power over polar radius and over angle. Power over radius corresponds to spatial frequency content of the image texture and that over angle reflects the orientations of the textural components. They use a distance metric to evaluate similarity of a query target to library images. Another spectral histogram method for texture classification is proposed by Xiuwen [11] that decomposes the spectral content of image textures using Gabor wavelets. These methods treat orientation of textural components as important to classification; hence make no attempt to impose rotational invariance on the representation.

Nill and Bouzas [12] propose a measure of image quality derived from the 2D spectrum of the image under evaluation. While their quality metric is based on the 2D image spectrum, these researchers present a 1D representation of the spectrum that has interesting properties in its own right. Nill and Bouzas employ 1D summaries of 2D power spectra to aid in visualizing the most significant spectral differences among images. The present investigation explores the use of this 1D spectral summary to test validity of fingerprint images directly and to diagnose acquisition problems such as aberrant sampling or faulty segmentation.

# 3. SIVV METHOD

## 3.1 IMAGE WINDOWING

The effect of observing the image signal over finite extent with discontinuity at the edges is to introduce spurious power into the spectrum (spectral leakage). Windowing reduces this effect, narrowing the peak side lobes and concentrating energy at the appropriate spatial frequency, thus better defining the location of peaks and resolving separate peaks that otherwise might remain merged. In order to enhance resolution and localization of spectral features, recommended procedure (see [13]) calls for applying a weighting function to the input signal prior to application of the Discrete Fourier Transform (DFT) such that values of the input signal taper toward zero at the edges. In the case of single fingerprint images, windowing has the additional advantage in reducing the effects of sensor contact artifacts at the sensor edge.

A variety of window functions were applied to several fingerprint image examples and the resultant spectra examined by inspection with respect to spatial frequency specificity of peaks as indicated by peak height, width, and number of distinct peaks. Such limited testing could not support exclusion of alternate window shapes, but the Blackman window seemed to produce reasonably well-defined peaks. Hence it was selected for the SIVV tool implementation. The 1D Blackman window is defined [14] as

$$w(n) = 0.42 - 0.5\cos\left(\frac{2\pi n}{N}\right) + 0.08\cos\left(\frac{4\pi n}{N}\right), \quad 0 \leq n \leq N. \quad (1)$$

The desired length of the window $L = N+1$ (or $L = M+1$) for windowing rows or columns of the image[i]. Conveniently, the application of this 1D window function to the input image requires only element-wise multiplication of the image rows and columns by 1D window functions of the appropriate length. Thus, first each row of the input image, $I(i,j)$, is multiplied by the window function $w(m), m = 1...M$ where $M$ = number of image columns. The columns of the output of this operation are then multiplied element wise by $w(n), n = 1...N$ where $N$ = number of rows of the image. Alternately the entire 2D window function, $W(n,m)$, may be constructed by forming the cross product of the column vector, $w(n)$, and the row vector $w(m)$. This cross product yields the 2D window function of the appropriate dimensions for element-wise multiplication by the input image, $I$.

---

[i] Square pixels are assumed here.

Fig. 1 illustrates the 2-dimensional Blackman window sized appropriately for application to the input image shown in fig. 2. The windowed image is shown on right of fig. 2 illustrating the tapering of intensity values toward the image edges. Various window functions share the property that they taper the image values toward zero on the borders.

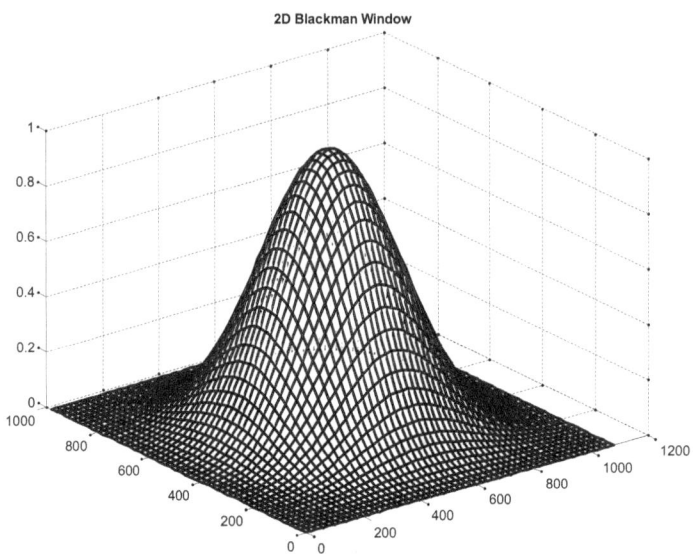

Fig. 1 2D Blackman window function.

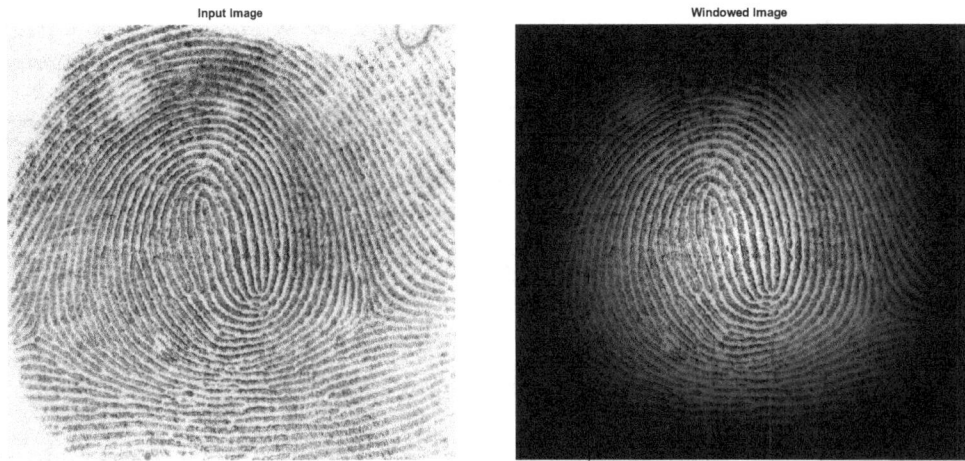

Fig. 2 Input image (995 x 1090 pixels) and Blackman windowed version.

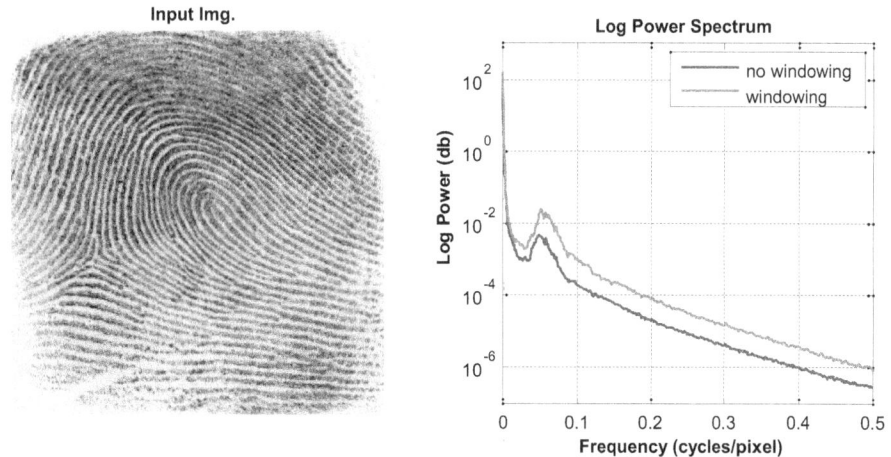

Fig. 3 Sample fingerprint image (922 x 860 pixels) and associated 1D spectra computed without and with application of a window function.

Fig 3 shows an example of spectrum computation without windowing and with application of the Blackman window to the image. Windowing increases the sharpness of the main peak at around 0.05 cycles/pixel and increases the power of the spectrum overall. A slight shift in frequency of the peak reflects the differential weighting applied to the central ridge structure relative to that on the periphery of the fingerprint. In this regard, it may be noted that the effect of the windowing will be at maximum for textures centered in the image frame. Effects of fingerprint offset will be examined later in this paper.

## 3.2 DISCRETE FOURIER TRANSFORM (DFT)

Given a 2D image, having dimensions $N$ rows x $M$ columns, a pixel grey level may be specified as $h(x,y)$ with $x$ coordinates ranging $0 \ldots M-1$ and $y$ coordinates $0 \ldots N-1$. The discrete Fourier transform (DFT), is computed as

$$H(u,v) = \sum_{x,y=0}^{M,N-1} \exp\left[2\pi i y \frac{v}{N}\right]\left[\exp 2\pi i x \frac{u}{M}\right] h(x,y) \quad (2)$$

$$u,v = -\frac{M,N}{2,2} \ldots \frac{M,N}{2,2}.$$

where $u$ and $v$ denote frequency components in $x$ and $y$ directions.

## 3.3 NORMALIZED POWER SPECTRUM

The 2D power spectrum is computed as

$$P_{(x,y)} = \frac{|H_{(x,y)}|^2}{|H(0,0)|^2} \quad (3)$$

The normalization term in the denominator is the square of the average grey level which corresponds to the dc term of the spectrum, $H(0,0)$.

In the output array of the Fast Fourier Transform (FFT) routine, the $0^{th}$ frequency term is repeated in each corner of the image frame and the spectra are mirrored about the middle rows and columns. Therefore, the 2D power spectral array must be rearranged to center the dc response, $P(0,0)$, in the image frame. For the 2D DFT this means that the quadrants are swapped diagonally such that quadrant 1 is swapped with quadrant 3 and quadrant 2 is swapped with quadrant 4 as shown in fig. 4.

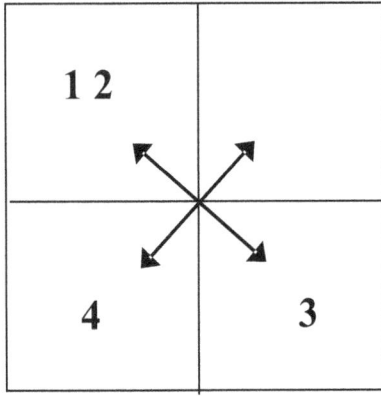

**Fig. 4 Diagram showing operation performed by a function that shifts quadrants of the DFT such that dc term, $P(0,0)$, is centered in the image.**

## 3.4 POLAR TRANSFORM OF POWER SPECTRUM

The polar transform is applied to the 2D power spectrum, converting rectangular coordinates of the 2D power spectrum to radius, $\rho$, and angle, $\theta$, with respect to the origin (center) of the 2D spectrum, where

$$\rho = \frac{\sqrt{u^2+v^2}}{\sqrt{M^2+N^2}} \quad (4)$$

$$\theta = \tan^{-1}\left(\frac{v}{u}\right) \quad (5)$$

Equation 3 then becomes

$$P(\rho,\theta) = \frac{|H(\rho,\theta)|^2}{|H(0,0)|^2} \quad . (6)$$

Values of $\theta$ range from 0 - 180°[ii] and radius values, $\rho$, correspond to frequencies 0 – N/2 cycles per image width (or height), where N is the minimum dimension of the input image. Noting that the maximum frequency that can be represented in an image requires 2 pixels, the spatial frequency scale is expressed as cycles per pixel from 0 to 0.5 cycles/pixel in increments of 0.5/(n-1), where n=number of radii specified for the polar transform.

## 3.5 1D NORMALIZED POWER SPECTRUM

To form a 1D representation of the polar transformed power spectrum one need simply sum over angle for each of the frequency (radius) values, i.e.

$$P(\rho) = \sum_{\theta=0}^{180} P(\rho,\theta), \quad 0...0.5 \; cycles/pixel \quad (7)$$

Rather than performing the normalization step as described in section 3.3, the present software implementation performs the DC level normalization here as

$$P(\rho) = \frac{P(\rho)}{P(0)}, \quad 0...0.5 \; cycles/pixel \quad (8)$$

Note that the radius, $\rho$, is expressed in units of cycles per pixel. For application of the polar transform, the number of radii is selected nominally as half the minimum of the length and width of the input image. Thus the polar transform of the 2D spectrum includes frequencies enclosed by the largest circle that can fit within the dimensions of the 2D spectral array.

In fig. 5, the 2D power spectrum is shown (top right) in image format. The units of frequency, $u$ and $v$, correspond to cycles per image dimension. The log function is applied to the magnitude values of this representation in order that small values at high frequencies might be displayed as well as the much larger power at low frequencies. The polar transform of the 2D spectrum is shown in fig. 5 (lower left). Radius values, $\rho$, represent cycles per image width corresponding to units of the 2D spectrum. Angles, $\theta$, in degrees, range from 0 to 180. The 1D spectral signal represented in the lower right of fig 5 is formed by summing over all angles the power at each radius of the polar transform. In the 1D spectrum, the radius units in cycles per image dimension are rescaled to the frequency range (0.0 to 0.5) cycles per pixel as described above.

---

[ii] As the spectrum repeats over the 360°, it is completely described over the interval 0-180°.

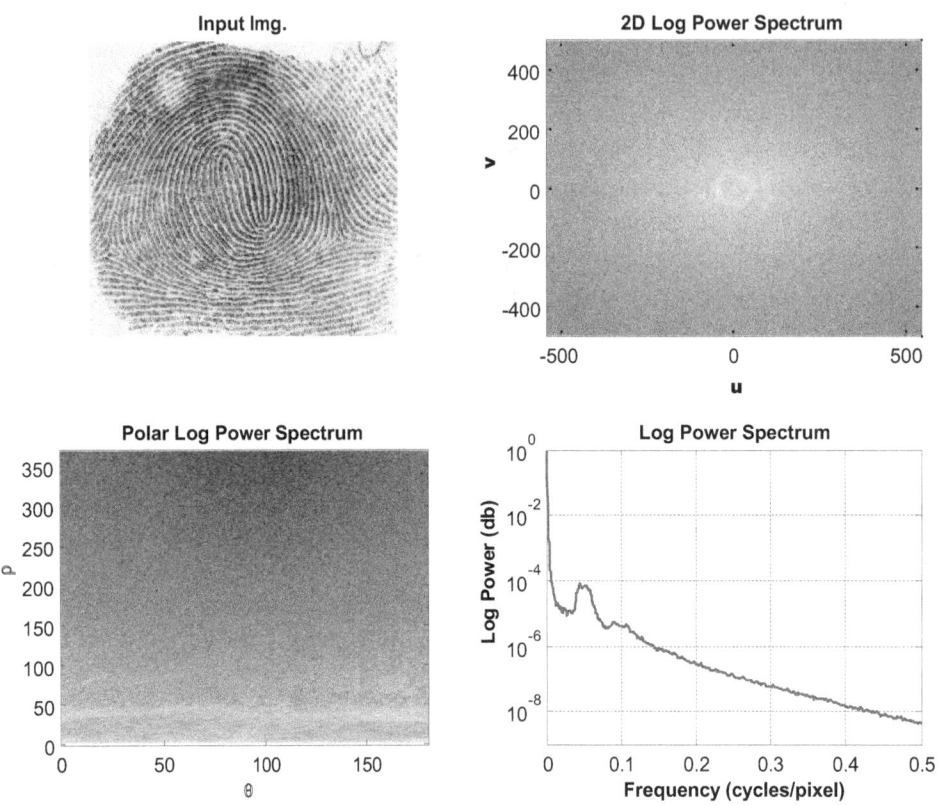

Fig. 5 Illustration of processing steps from image to 1D Normalized Power Spectrum. (Image of size 995 x 1090 pixel) Spatial frequency, $u$ and $v$, are in units of cycles per image width as are the unnormalized units of polar radius, $\rho$. Polar angles, $\theta$, are in degrees. In log power spectrum, radius values of cycles/image width are scaled to become cycles/pixel.

It is noted here that units of cycles per pixel may be converted to a more conventional spatial frequency specification given that information is available by which to assign a conventional length unit to pixel dimensions. For many images this length factor may remain unknown, or recoverable only through estimation based on assumptions relative to the image content. For fingerprint images, knowledge of the sample rate often is available in metadata supplied with the imagery, and required to be conformant to the ANSI/NIST standard (see [15]) for the Type 1 fingerprint data record. This information enables assignment of a conventional linear dimension, e.g. millimeters, to the pixel. The generic unit, cycles per pixel, is converted to cycles per millimeter by multiplying by the appropriate scaling factor. However, for image applications, where in most cases scale is unknown, it could be argued that cycles per pixel is a perfectly acceptable unit of frequency.

As may be observed via inspection of a variety of examples, the 1D spectral representation of fingerprints appears consistently to exhibit a distinctive valley-peak combination (as observed in the lower right plot of fig. 5) in the low frequency range, doubtless the expression of the characteristic friction-ridge flow structure of the fingerprint, the so-called level 1 detail.

## 3.6 PEAK STRUCTURE OF SPECTRUM

The feature of potential diagnostic significance is observed in the appearance of paired local minima and maxima of power values in a frequency band related to the ridge spacing. Toward further development of this notion as a diagnostic, an algorithm is devised to locate peaks and valleys in the spectrum and to select pairs of such features that might be examined further with respect to position along the frequency axis, height of the peak relative to preceding valley ($dy$), frequency distance between valley and peak ($dx$), slope ($dy/dx$), and statistics of these values with respect to image classes. Other features of potential value may be the number of peaks detected in the spectrum and the ordinal position of the valley-peak pair having the maximum value of $dy$. Further, total or average power in each of a small set of frequency bands might also be taken as diagnostic features.

Figs 6 - 8 exhibit fingerprint images and their spectra. Red and green dots along the spectrum denote local minimum and maximum feature pairs having the largest power difference, i.e., $dy$ value. The feature definition requires that the peak be preceded (over increasing frequency) by a local minimum.

In the present implementation, the spectrum may be smoothed (optionally) prior to peak localization using a $n$-point moving average filter, where $n$ is an odd number greater than 1. In the examples $n=7$. The filtering algorithm performs zero-phase digital filtering by processing the input data in both forward and reverse directions as described in [16].

The peak finder returns coordinates of pairs of local minimum and maximum values. These are processed further to yield the pair exhibiting the maximum difference in power, i.e. $dy$. The frequency position of the feature pair for fingerprints on average should be related to the ridge spacing, and in most cases the spectra for fingerprints are expected to lack other significant peak structure. This aspect is thus proposed as a means to validate fingerprint images.

The feature pattern occurs over a variety of fingerprint formats. Even the more complicated slap-four print exhibits the structure rather clearly as shown in fig. 8.

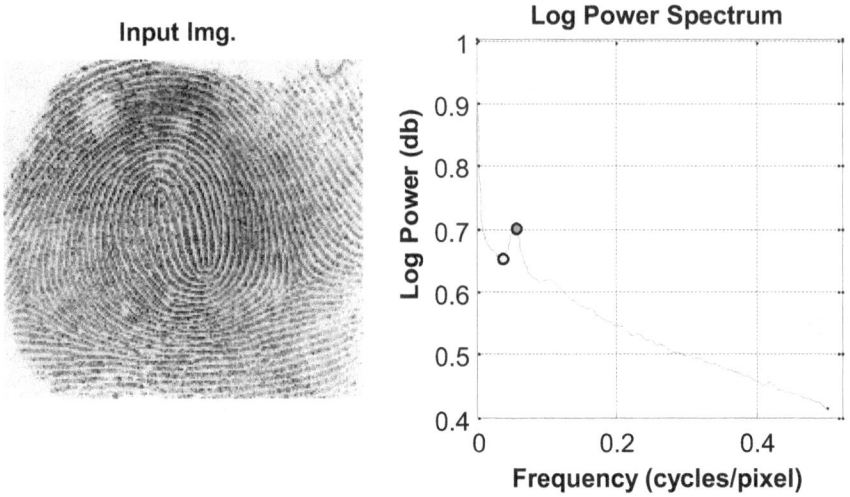

Fig. 6 Fingerprint sampled at 1000 ppi (1107 x 1145 pixels) and associated 1D spectrum. The pair of local minimum and maximum having the largest difference in power are indicated with green and red marks.

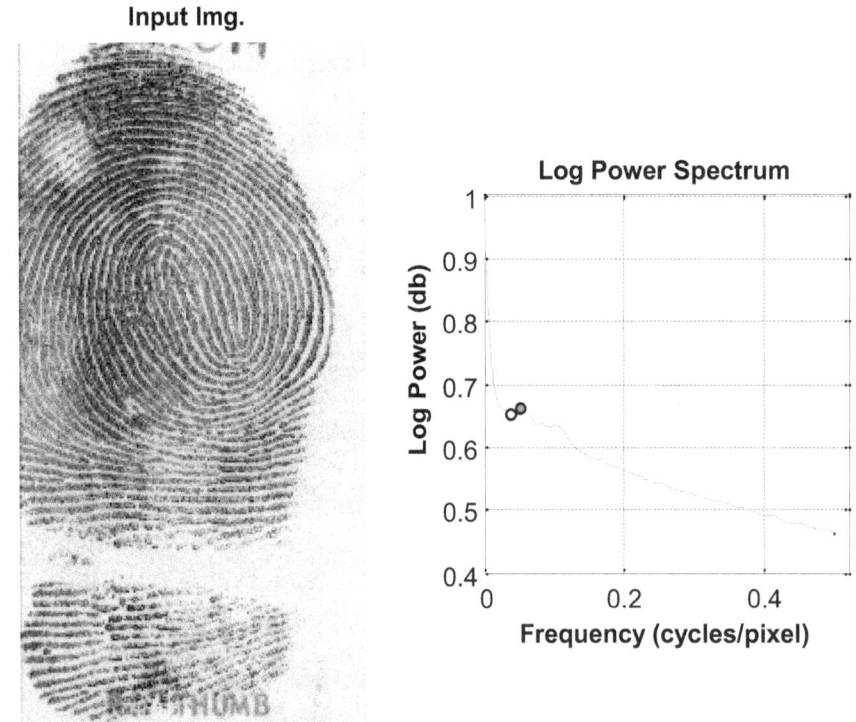

Fig. 7 Fingerprint sampled at 1000 ppi (1990 x 790 pixels) and associated 1D spectrum.

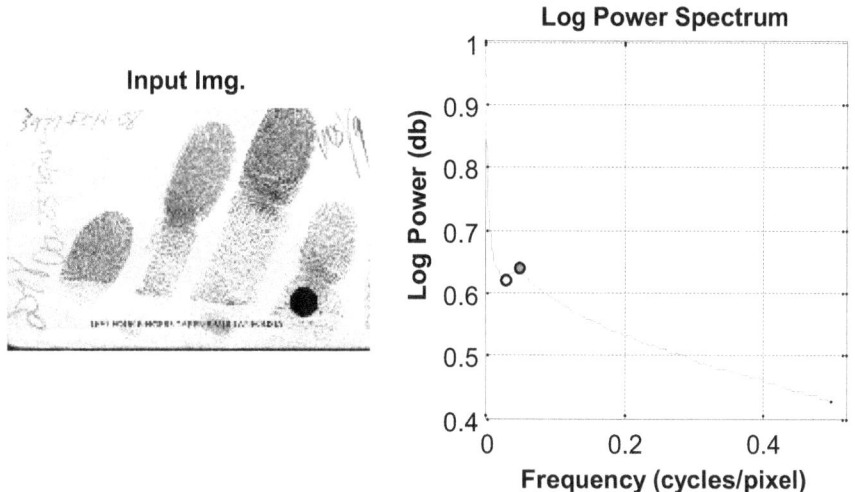

Fig. 8 Slap-four image sampled at 1000 ppi (2049 x 3116) and spectrum.

# 4. RESPONSE OF THE SIGNAL TO IMAGE PARAMETERS

## 4.1 ROTATION INVARIANCE

A demonstration of the rotation invariance of the SIVV signal is provided in Appendix A. But rotation invariance is intrinsic to the signal representation itself. The 2D spectrum is not rotation invariant in that the 2D components preserve directionality of frequency information in the image. Orientation of frequency components is conveyed to the polar transform of the 2D spectrum in which rotation becomes translation. Thus, polar transformed spectra of images rotated with respect to each other may be brought into registration by circular shift [iii] along the angular dimension. The 1D spectral summary used in the SIVV utility, however, is rotation invariant without additional manipulation. That is, integration of the polar spectrum over angle generates a similar 1D representation regardless of where one places the origin of the summation operation.

## 4.2 RESPONSE OF SPECTRUM TO TRANSLATION

Whereas the SIVV signal representation is invariant to rotation of the fingerprint sample, the same is not true for translation of the main ridge-flow region of the fingerprint from the image center. We find that fingerprint position does not affect the frequency structure of the power spectrum. However, with windowing, the power is substantially greater with the centered fingerprint. We find further that "white" space surrounding the fingerprint also tends to reduce the spectral power. Fortunately, both potential problems are solved easily as will be explained.

To examine the effect of fingerprint offset from the frame center, a test image displaying the fingerprint in a non-centered position was selected from an available database of images scanned at 1000 pixels per inch (ppi). A second image was created from this image by shifting the fingerprint from its placement along one edge toward the center of the frame by applying a circular shift to the image. Figs 9 and 10 exhibit the effects of this manipulation on the spectrum both with and without pre-DFT application of a weighting window.

---

[iii] Columns of pixels are removed from one edge of the array and inserted along the opposite edge. Some number of such operations should bring polar transform arrays of rotated images into correspondence.

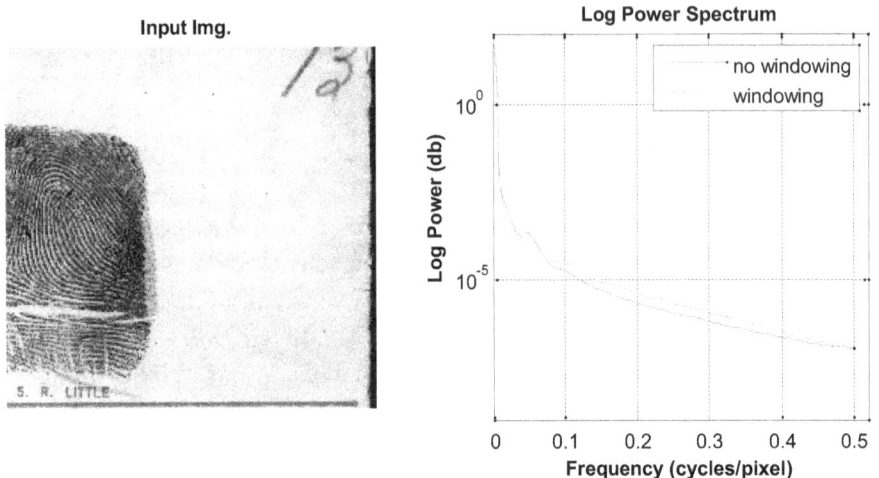

**Fig. 9 Off-center fingerprint image (1500 x 1600) and associated spectra computed both without and with Blackman windowing.**

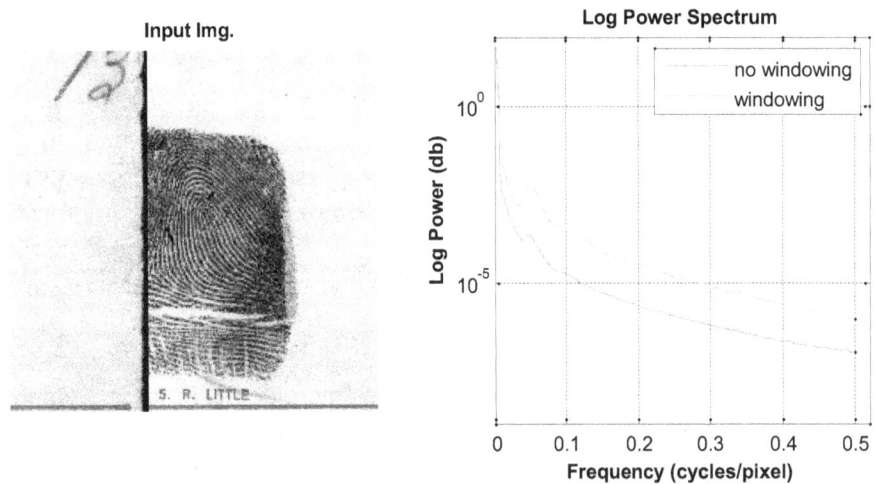

**Fig. 10 Image from fig 9 with fingerprint moved toward more centered position applying a circular shift to the image.**

The spectra of figs 9 and 10 confirm that the position of the fingerprint in the image has only negligible effect on the position of the peaks in the spectrum. Moreover, power is almost identical for centered and offset fingerprints in the "no windowing" case. Significant increase in the power is seen with application of the Blackman window to the centered fingerprint as shown in fig 10. Windowing has little effect on the spectrum of the offset fingerprint shown in fig 9. Thus, while fingerprint position has little effect on the pattern of spectral peaks with respect to frequency localization, the power and definition of the peaks can be increased by centering the fingerprint in the frame prior to application of the weighting window.

The latter result suggests advantages and perhaps even the necessity of centering the fingerprint in the image before estimating the spectrum, particularly if windowing is to be employed. The spectral power response is maximized by centering the fingerprint in the image frame. Also, we find that the spectral power is enhanced further and the signature made more consistent by excluding empty space surrounding the fingerprint. Sophisticated image segmentation schemes might be used to center and crop the fingerprint from surrounding white space, but simple, less computationally expensive methods are quite effective.

For example, in the current **MATLAB**[iv,v] implementation of the SIVV, the average row and column coordinates of the binary output of a Sobel [17] edge filter are taken as the center of the fingerprint. A circular shift in row and column is then performed to shift the centroid to the geometric center of the image prior to windowing and spectrum computation. Except in cases having significant stray marks in the surrounding of the fingerprint, this method does a good job of centering the ridge-flow of the image. Where edge "noise" is significant in the background, cropping is necessary. The same edge map can be used fro cropping as well as for centering.

In the C++ implementation of the SIVV method to be detailed in a subsequent document, both centering and cropping functions are implemented using the output of a Canny[18] edge detector. The fingerprint center is determined as described above. Then row and column cropping boundaries are set by examining edge point density statistics along rows and columns with respect to a threshold density value. Only the image region bounded by edge densities above threshold is processed further.

Thus, the sensitivity of the SIVV signal to translation and background is corrected easily. Yet this sensitivity of the SIVV method to translation points to the additional diagnostic capacity of the method to detect fingerprints offset from the ideal centered position. Difference between power spectra before and after centering operations would indicate that the original fingerprint was not centered in the original image frame and provide an estimate of the degree of offset.

---

[iv] Registered trademark of The Mathworks, Inc. 3 Apple Hill Drive Natick, MA 01760-2098.

[v] Any mention of commercial products within this report is for information only; it does not imply recommendation or endorsement by NIST

## 4.3 SCAN SAMPLE RATE SENSITIVITY (1000 PPI VS. 500 PPI)

Fingerprints, typically, have been scanned at a sample rate of 500 ppi. With improvements in scanner technology, data storage capacity, processing capability, and growing interest in extending analysis to smaller features of the fingerprint, movement has been rapid toward scaning at higher rates of 1000 ppi and even up to 2000 ppi. Accordingly, the potential for matching failures due to an unrecognized disparity between image sample rates has become significant. Verification of image sample rate, accordingly, has become an important component of database screening.

Given that the 1D spectral frequency is scaled as cycles per pixel, one might expect that reducing the sample rate by a factor of two should expand the frequency representation by a proportional degree and shift the peak toward higher frequency. Indeed this occurs and may be useful for detecting large differences in scan sample rate. Given some variation in the precise frequency position of the valley-peak feature among fingerprints at a constant scan sample rate, the spectrum may not be sufficiently sensitive to detect small differences in sample rate. However, the method can signal disparities on the order of several hundreds of ppi.

Identical fingerprints sampled at both 1000 ppi and 500 ppi were not available, so the comparison images are prepared by resampling a 1000 ppi fingerprint at half the original rate. The software function employed for this operation applies an "anti-aliasing" (low-pass) filter to the image prior to resampling the image. This ensures that under-sampling the high frequency components of the original image will not create spurious low frequencies, i.e., the Nyquist sampling criterion [19, 20] will be satisfied. Resampling was done using a bicubic interpolation method.

Example 1000 ppi and 500 ppi images and their spectra are shown in figs. 11 and 12. As expected, the structure of the spectrum is essentially the same in the two cases but the features are stretched out, i.e. redistributed, over the frequency dimension for the lower sample rate. At the half the sampling rate, the spectrum at 500 ppi covers only ½ of the actual frequency range of that of the 1000 ppi sampling. Hence a 500 ppi image should be easy to recognize among 1000 ppi images.

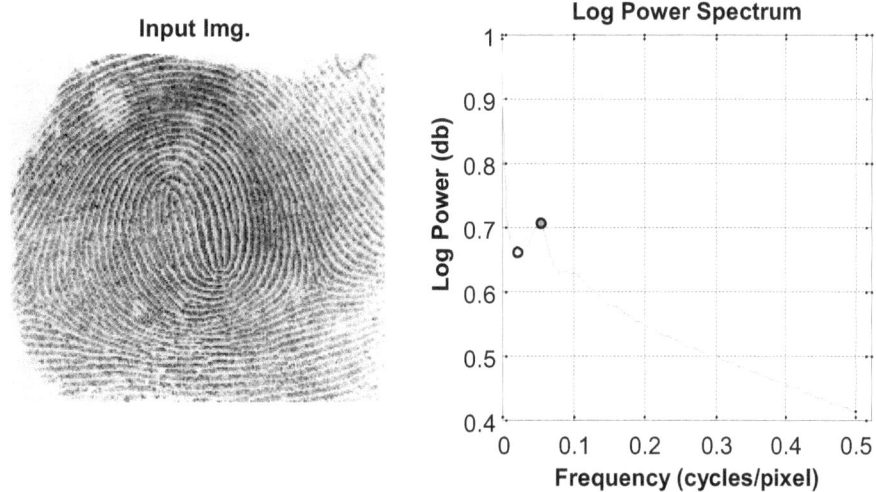

Fig. 11 Fingerprint image digitized at 1000 ppi (995 x 1090 pixels).

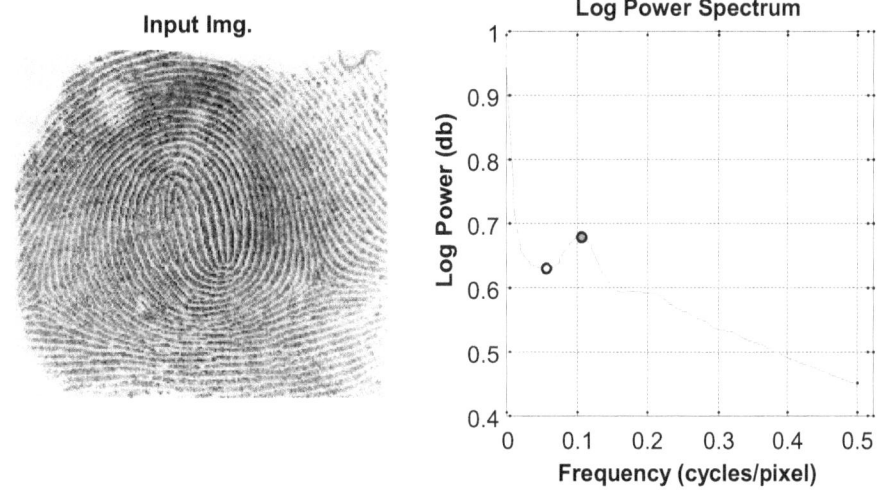

Fig. 12 Fingerprint image from fig. 11 resampled at 500 ppi (498 x 545 pixels) using bicubic interpolation. The spectral features are shifted by factor of 2 in frequency.

## 5. RESULTS AND DISCUSSION

The SIVV code was run on each of the mixed datasets (described in Appendix B) and the output piped to a text file for analysis. As indicated elsewhere in this document, the current output of the SIVV consists of the following:
- image file name
- ordinal location of the maximum peak among the array of peaks returned by the peak finder
- number of peaks returned by the peak detector
- power difference ($dy$) between the maximum peak and the signal minimum (valley) immediately preceding it
- frequency difference ($dx$) between the valley and peak
- slope between the valley and peak ($dy/dx$)
- frequency of the midpoint[vi] between the valley and the peak

The output of the SIVV routine is examined for each of the datasets with respect to the distributions of the feature values among fingers in the case of the fingerprint databases (see Appendix C), and with respect to fingerprint compared to non-fingerprint images. Then assessment is made of the error rates in classifying images as fingerprint or non-fingerprint considering the peak height relative to the valley ($dy$) feature with a rough constraint on the acceptable frequency location of the valley-peak feature.

Figs 13 and 14 contrast distributions of peak height and feature frequency location for fingerprints and non-fingerprint images of the mixed dataset. Fig. 13 shows fingerprint images to exhibit significantly higher peak height values in comparison to non-fingerprint images. Thus, this feature appears effective at discriminating fingerprints from other images.

---

[vi] Initial analysis considered the diagnostic feature to consist of both a minimum and a succeeding maximum. Given this feature definition, the midpoint seemed a reasonable means to locate the feature. Currently, however, we are considering the frequency position of the peak alone as a feature, particularly in differentiating various sample scan rates, and have plans to test such.

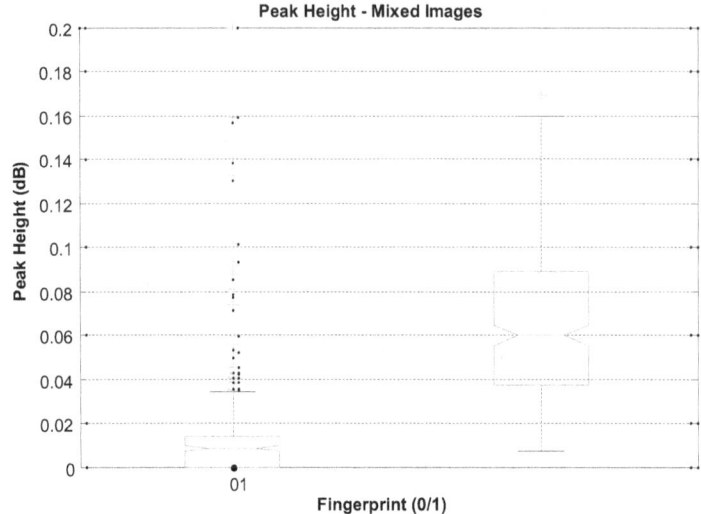

Fig. 13 Box plot showing distributions of peak height for non-fingerprint images (0) and fingerprint images (1) in a mixed image dataset.

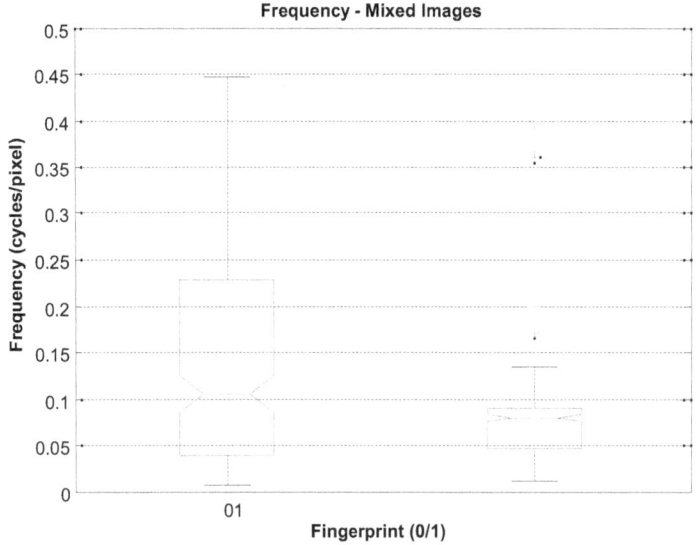

Fig. 14 Box plot showing distributions of frequency location of the valley-peak structure for non-fingerprint images (0) and fingerprint images (1) in a mixed image dataset.

Frequency location is less discriminating, though the range of frequency location is much smaller for fingerprints than for non-fingerprint images as is shown in fig. 14. This suggests that frequency location may not be particularly diagnostic in its own right but rather may provide a constraint or filter in selection of peak height values for classification. Fig 15 shows the relative error rates for fingerprint vs. non-fingerprint classification fixing a maximum frequency location at 0.15 cycles/pixel and varying the threshold of peak height. An equal proportion of false positives and false negatives, equal error rate (EER), of approximately 10 % is achieved at a peak height ($dy$) threshold of around 0.024 dB.

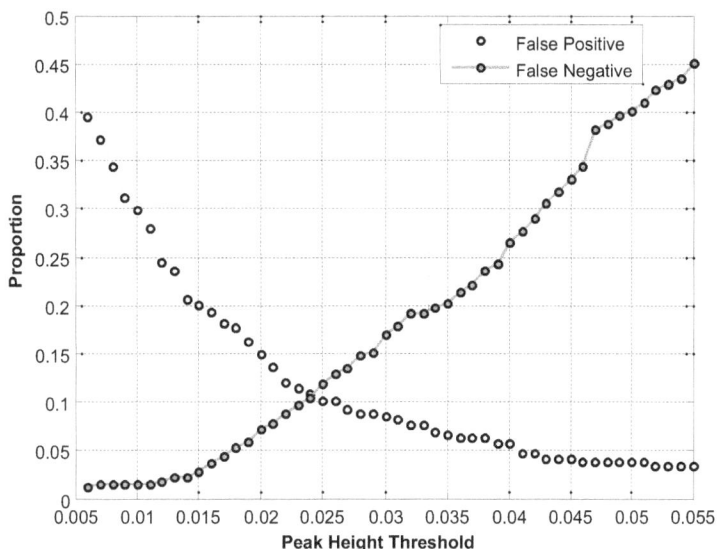

Fig. 15 Plot of false positive vs. false negative error rates with variation of threshold value of peak height used as a classifier applied to a mixed image dataset. Feature frequency maximum was set at 0.15 cycles/pixel.

Figs. 16 and 17 exhibit distributions of peak height and feature frequency location for fingerprint, face, and iris images of the mixed biometric dataset, Mixed Biometric 1. As with the general mixed image dataset, the distribution of peak heights of fingerprints are quite distinct from the much lower values for face and iris images. Again, frequency location is not unique for fingerprints, but may serve as a limiting filter for selection of peak height as a classification criterion.

The composite consideration of both peak height and frequency is illustrated in the scatter plot of fig. 18. The fingerprint responses are largely separated from those of both face and iris images. Moreover, separate clusters are evident for fingerprints at the two scan sample rates, 1000 ppi and 500 ppi. Face images represented within the fingerprint clusters are actually very small area faces against a brick wall background. Iris images among the fingerprints exhibit prominent eye lashes and eyebrows generating the periodic structure.

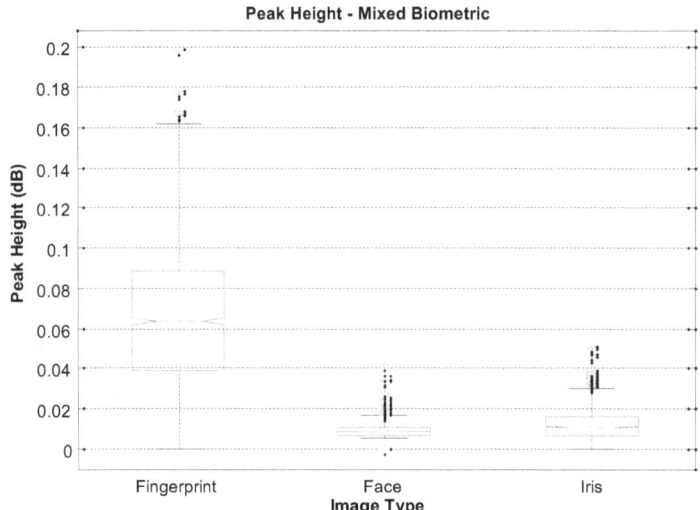

Fig. 16 Box plot showing distributions of peak height ($dy$) for equal numbers of fingerprint, face, and iris images in a mixed biometric dataset.

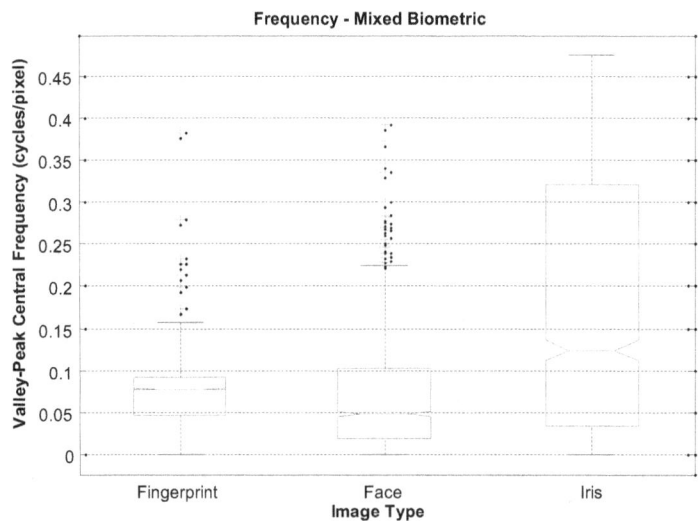

Fig. 17 Box plot showing distributions of frequency location of the valley-peak feature for equal numbers of fingerprint, face, and iris images in a mixed biometric dataset.

Fig. 19 shows the error rate as a function of threshold in peak height for classification of fingerprint vs. non-fingerprint (face or iris). Again the upper frequency limit is set at 0.15 cycles/pixel. An EER of 6.6 % is achieved at a peak height of 0.02 dB.

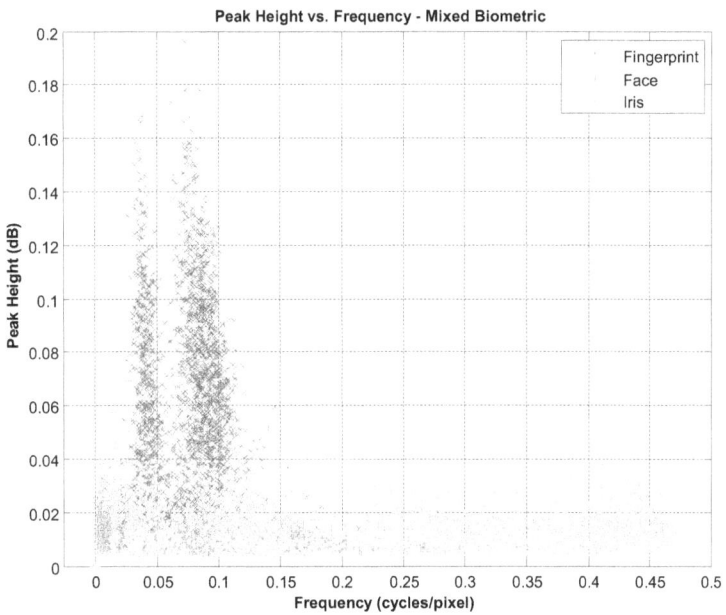

Fig. 18 Peak height ($dy$) vs. frequency of SIVV output for images of the Mixed Biometric 1 dataset. Note the separation of fingerprints into clusters corresponding to the 1000 ppi and 500 ppi samples.

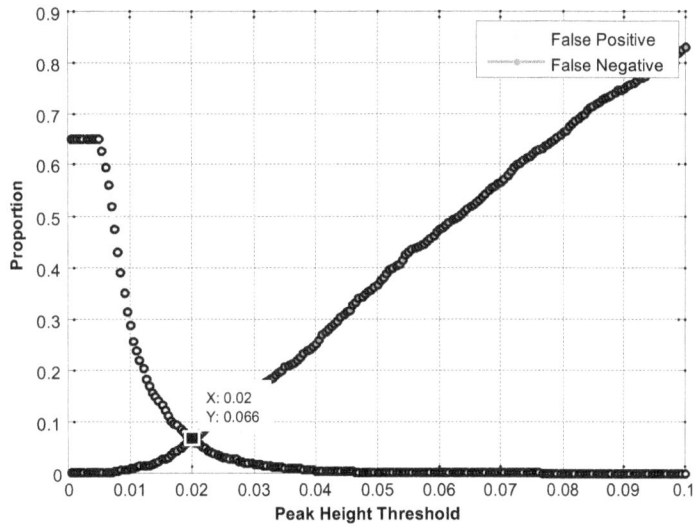

Fig.19 Plot of false positive vs. false negative error rates with variation of threshold value of peak height used as a classifier applied to the mixed biometric dataset. Feature frequency maximum was set at 0.15 cycles/pixel.

In the preceding EER analysis only the peak height feature is used for classification, though constrained via a maximum limit on frequency position. From examination of the distributions of all the features, however, it is not clear that consideration of the other features in a simple fashion will be adequate to enhance the classification performance beyond that reported here. Training a support vector machine (SVM) or other learning model would yield a more sophisticated integration of the multiple features. However, it is also possible that a more precise representation of the spectral shape, i.e. beyond merely the major peak, would provide greater classification accuracy. That is, examination of some of the non-fingerprint images that exhibited a peak feature similar to that of the fingerprint also exhibited peaks at higher frequencies that would differentiate them from fingerprints. Hence a follow-on effort will examine the spectrum itself in greater detail via additional features representing average or total power in each of some small number of frequency bands of the spectrum in addition to the features currently output. For example, the frequency extending from 0.0 to 0.5 cycles/pixel might be examined over each of 10 intervals of 0.05 cycles/pixel in width. A classification model based on these additional 10 features together with those already discussed might improve the classification accuracy.

## 6. CONCLUSIONS

The spectral image validation and verification (SIVV) method proposed here demonstrates its utility as a means for screening fingerprint databases for poorly segmented specimens and for non-fingerprint images that may have been included in the database. The magnitude of the distinctive spectral feature related directly to the level 1 ridge flow feature of the fingerprint provides a primary diagnostic indicator of the presence of a fingerprint image. The frequency location of the feature provides a secondary classification metric and on a coarse level verifies the reported scan sample rate of the fingerprint image. Using only a simple threshold criterion based on these two features, the SIVV utility was able to limit classification EER to 10% or less for a mixed image dataset, including many images intentionally selected to confuse the algorithm, and a dataset including fingerprints amidst face and iris images. Efforts toward improving performance may include use of a more sophisticated classification model using additional features available in the 1D power spectrum.

# 7. REFERENCES

[1] E. Tabassi, C. L. Wilson, C. I. Watson, "Fingerprint Image Quality: NISTIR 7151:," National Institute of Standards and Technology, Gaithersburg, NISTIR 08/19/2004 2004.

[2] E. Tabassi and C. L. Wilson, "A novel approach to fingerprint image quality," in *Image Processing, 2005. ICIP 2005. IEEE International Conference on*, 2005, pp. II-40.

[3] F. Alonso-Fernandez, J. Fierrez, J. Ortega-Garcia, J. Gonzalez-Rodriguez, H. Fronthaler, K. Kollreider, and J. Bigun, "A Comparative Study of Fingerprint Image-Quality Estimation Methods," *Information Forensics and Security, IEEE Transactions on,* vol. 2, pp. 734-743, 2007.

[4] M. S. Altarawneh, W. L. Woo, and S. S. Dlay, "Objective Fingerprint Image Quality Assessment using Gabor Spectrum Approach," in *Digital Signal Processing, 2007 15th International Conference on,* 2007, pp. 248-251.

[5] T. P. Chen, x. Jiang, and W. Y. Yau, "Fingerprint image quality analysis," in *Image Processing, 2004. ICIP '04. 2004 International Conference on,* pp. 24-27.

[6] H. Lin, W. Yifei, and A. Jain, "Fingerprint image enhancement: algorithm and performance evaluation," *Pattern Analysis and Machine Intelligence, IEEE Transactions on,* vol. 20, pp. 777-789, 1998.

[7] D. Maio and D. Maltoni, "Ridge-line density estimation in digital images," in *Pattern Recognition, 1998. Proceedings. Fourteenth International Conference on,* pp. 534-538.

[8] Y. Yin, J. Tian, and X. Yang, "Ridge distance estimation in fingerprint images: Algorithm and performance evaluation," *Eurasip Journal on Applied Signal Processing,* vol. 2004, pp. 495-502, 2004.

[9] E. Lim, J. Xudong, and Y. Weiyun, "Fingerprint quality and validity analysis," in *Image Processing. 2002. Proceedings. 2002 International Conference on,* 2002, pp. I-469.

[10] C.-j. Wang, Y.-b. Yang, W.-j. Li, and S.-f. Chen, "Image texture representation and retrieval based on power spectral histogram," in *Tools with Artificial Intelligence, 2004. ICTAI 2004. 16th IEEE International Conference on,* 2004, pp. 491-495.

[11] L. Xiuwen and W. DeLiang, "Texture classification using spectral histograms," *Image Processing, IEEE Transactions on,* vol. 12, pp. 661-670, 2003.

[12] N. B. a. B. Nill, B. H., "Objective Image Quality Measure Derived from Digital Image Power Spectra," *Optical Engineering,* vol. 31, pp. 813-825, 1992.

[13] F. J. Harris, "On the use of windows for harmonic analysis with the discrete Fourier transform," *Proceedings of the IEEE,* vol. 66, pp. 51-83, 1978.

[14] A. V. a. R. W. S. Oppenheim, "Discrete-Time Signal Processing," Upper Saddle River, NJ: Prentice-Hall, 1999, pp. pp. 468-471.

[15] I. American National Standards Institute, *NIST Special Publication 500-275 - American National Standard for Information Systems -- Data Format for the Interchange of Fingerprint, Facial, & Other Biometric Information - Part 2: XML Version*. Gaithersburg, MD: U.S. DOC/National Institute of Standards and Technology, 2008.

[16] F. Gustafsson, "Determining the initial states in forward-backward filtering," *IEEE Transactions on Signal Processing,* vol. 44, pp. pp.988–992, April 1996, 1996.

[17] I. Sobel, and G. Feldman, "A 3x3 Isotropic Gradient Operator for Image Processing, presented talk at the Stanford Artificial Project 1968, unpublished," in *Pattern Classification and Scene Analysis*, R. a. P. H. Duda, Ed. Menlo Park, CA: John Wiley & Sons, 1973, p. 482.

[18] J. Canny, "A Computational Approach to Edge Detection," *Pattern Analysis and Machine Intelligence, IEEE Transactions on,* vol. PAMI-8, pp. 679-698, 1986.

[19] C. E. Shannon, "Communication in the Presence of Noise (Classic Paper Reprint) " *Proceedings of the IEEE,* vol. 86, pp. 447-457, February 1998 1998.

[20] M. Unser, "Sampling - 50 Years After Shannon," *Proceedings of the IEEE,* vol. 88, pp. 569-587, April 2000 2000.

[21] M. Frigge, David C. Hoaglin, and Boris Iglewicz, "Some implementations of the Boxplot," *The American Statistician,* vol. 43, pp. 50-54, 1998.

[22] J. W. Tukey, *Exploratory Data Analysis*. Reading, MA: Addison-Wesley, 1977.

## 8. APPENDIX A: DEMONSTRATION OF ROTATION INVARIANCE

Experiments indicate the 1D spectrum to be largely invariant to in-plane rotation of the fingerprint. In order to avoid effects of cropping or scale change that would result from rotating a rectangular image, circular regions of interest are extracted from the original images. The fingerprint samples are rotated by resampling using bicubic interpolation. In the case of $90°$ rotation, the image transpose is used. Figs A-1 & A-2 demonstrate the stability of the 1D spectral representation over rotation for two images.

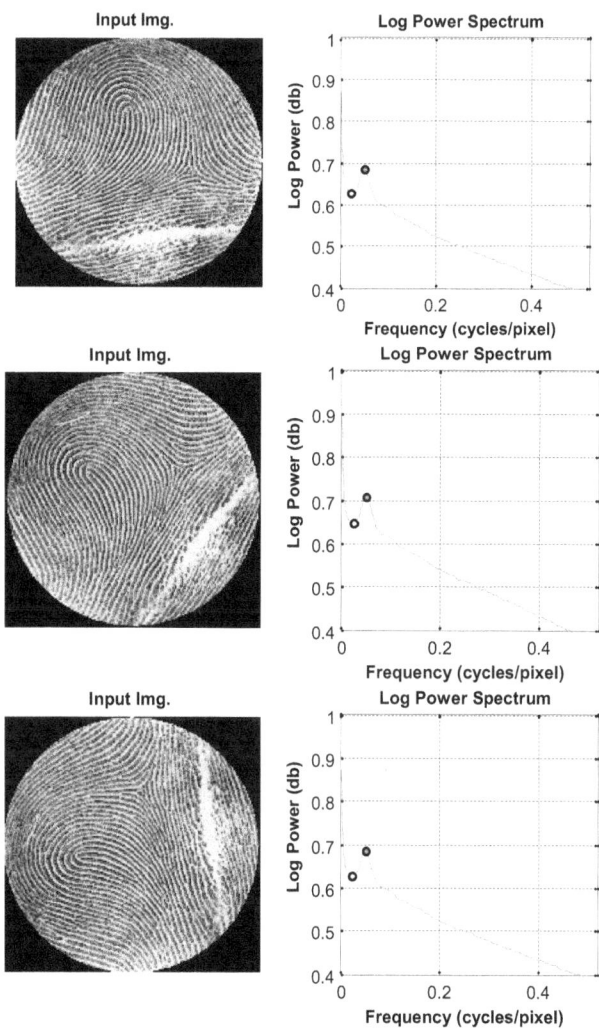

Fig. A-1 Circular sample of a rolled fingerprint image (1029 x 1023) captured at 1000 ppi. Original orientation (top) is rotated 45° and 90°.

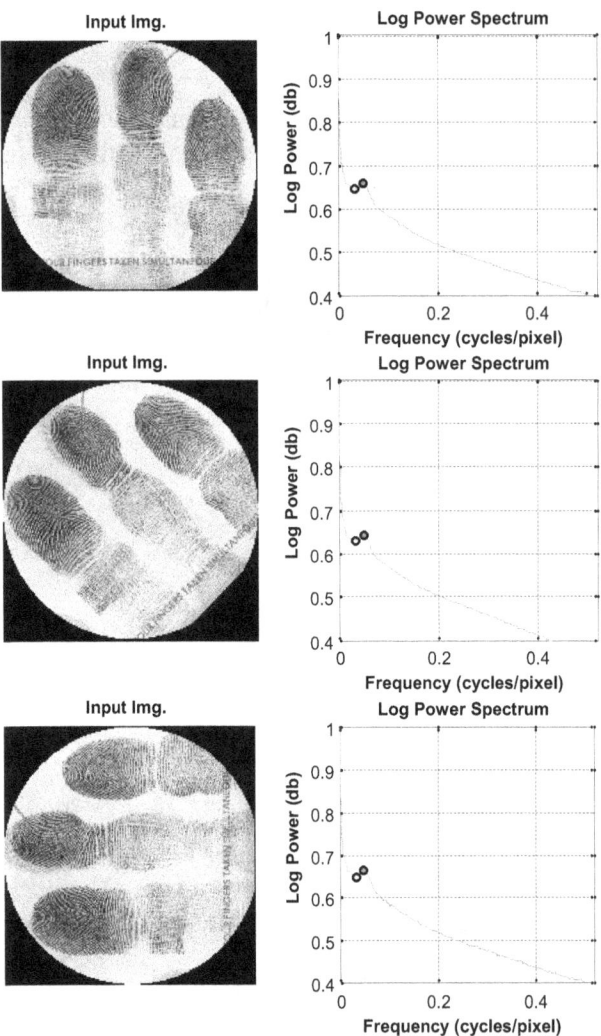

Fig. A-2 Circular sample of a slap-four image (2012 x 2064) captured at 1000 ppi. Original orientation (top) is rotated 45° and 90°.

# 9. APPENDIX B: TEST DATA

Performance of the software relative to the designed use of fingerprint image validation and verification (IV&V) is evaluated with respect to several fingerprint and non-fingerprint databases. The databases used are characterized as follows:

A fingerprint database, here referred to as SD27_1000, consists of images digitized at a sample rate of 1000 ppi from FBI standard 10-print cards. The individual images include rolled ink prints of each of the 10 fingers, plain impressions of each of the two thumbs, and plain "slap-four" impressions of the right and left hands. The dataset includes images from 212 fingerprint cards with 14 images per card for a total of 2968 images.

A second dataset is derived from this image set by resampling the 1000 ppi images to 500 ppi, with application of a low-pass filter to reduce the effects of aliasing due to under-sampling the original images. This dataset is referred to as SD27_500.

A third fingerprint dataset, referred to as SD29_Plain consists of 1188 plain impressions of fingers 1-10. These fingerprints were digitized at 500 ppi via scanning of the inked impressions.

The mixed data set referred to as ImagesTest02 consists of a variety of images largely downloaded from the Internet (e.g. Google Images collection) and various other public sources. The 331 downloaded images include a small number of biometric data samples such as fingerprints, iris, and face at various scales included in website libraries. Many non-biometric images in this sample were selected on the basis of their exhibiting periodic structure that might generate a spectrum similar to that of a fingerprint. Others images include portraits, trees, landscapes, buildings, textured materials, and other subjects at a variety of pixel resolutions.

ImagesTest02 contained a relatively small number of actual fingerprint images. Accordingly, the dataset was expanded for testing by adding a random sample of fingerprint images drawn from the three fingerprint datasets, SD27_1000, SD27_500, and SD29. Thus, to the original 331 images of the mixed set were added 300 additional images including 100 images drawn from each of the fingerprint datasets for a total of 631 images. We refer to this fourth dataset as ImagesTest03.

Finally, fifth and sixth datasets, Mixed Biometric1 and Mixed Biometric2, were constructed to examine relative distributions of the three biometrics and to examine classification rates for fingerprints against the background of non-fingerprint biometrics. A suite of face and iris images was obtained from Phillips and Scallan of NIST. These images are part of the dataset assembled for use in the Multi-Biometric Grand Challenge (MBGC)[vii] conducted by these researchers. The face images consist of frontal posed subjects imaged at a range of scales from full-frame face to images in which the subject is at a considerable distance from the camera with the face occupying only a very small area of a natural scene image. In a number of these images the comparatively small area face is imaged against a background often consisting of a brick wall. In some cases subjects are wearing striped or plaid clothing. The iris images are acquired using short wavelength infra-red (near-IR) lighting and include the entire eye, including eyelashes. In most of the images, the iris is nearly all visible or only slightly cropped. In many of the iris images, the eyebrow is visible.

Mixed Biometric 1 used to compare distributions of spectral features was constructed from equal sized random samples of face, iris, and fingerprint images (3000 of each). For the classification experiment, fingerprint vs. non-fingerprint, a random sample of 3000 fingerprint images was combined with 1500 each of face and iris images to form Mixed Biometric 2.

---

[vii] http://face.nist.gov/mbgc/

# 10. APPENDIX C: DISTRIBUTIONS OF SIVV FEATURES

Figs C-1 – C-3 exhibit box and whisker plots [21, 22] summarizing the distributions of the peak height ($dy$) values grouped by finger designation for the three fingerprint datasets. In each figure, the enclosed region, the box, includes observations within the $25^{th}$ and $75^{th}$ percentiles with the median ($50^{th}$ percentile) indicated by the red line. The vertical lines terminated by a short horizontal line extend for 1.5 times the lower and upper interquartile ranges. Observations beyond these intervals are considered to be outliers and are indicated by the red "plus" symbols. The notches centered on the medians indicate the 95 % confidence intervals of the medians and can be used to compare medians across the various boxes. For example, a median included within the notched region of another is not significantly different in the statistical sense, though one is advised to make such an interpretation with caution. In practice, we are more impressed with significant offset between boxes.

Figs C-1 and C-2 show respectively the peak height distributions for the 1000 ppi and 500 ppi images of the SD27 dataset. Within each of the figures, the distributions are similar among the rolled prints, having the designations 1 – 10, though heights are slightly higher for the lower sample rate. In both cases the peak heights are reduced significantly for the plain thumb impressions, 11 and 12, and for the slap-four plain impressions of 13 and 14. Inspection of images finds that in general the contrast is higher with the rolled prints, and the ridge pattern occupies a greater proportion of the image area than in the plain impressions. Moreover, handwritten content on many cards is placed over the plain print regions of the card, introducing frequency content in potential conflict with that of the ridge pattern.

The position of the valley-peak structure, taken as the midpoint between the valley minimum and the peak maximum, are also consistent among the rolled prints 1 – 10, with the bulk of the values falling between 0.03 cycles/pixel and 0.05 cycles/pixel for the 1000 ppi images (see fig. C-4) and between 0.07 cycles/pixel and 0.09 cycles/pixel for 500 ppi images as shown in figs C-5 and C-6. Inasmuch as the ridge frequency metric for each print represents an average over the entire print, and in fact, the entire image, a more precise position is not expected. The distribution for the plain single thumb prints, 11 and 12, exhibit much greater variability and a noticeable shift toward higher frequency relative to the distributions for the rolled prints. By contrast, the frequency position for the plain slap-four prints, 13 and 14, is distributed with smaller variance than the other distributions and the frequency position is slightly lower than that of the rolled prints.

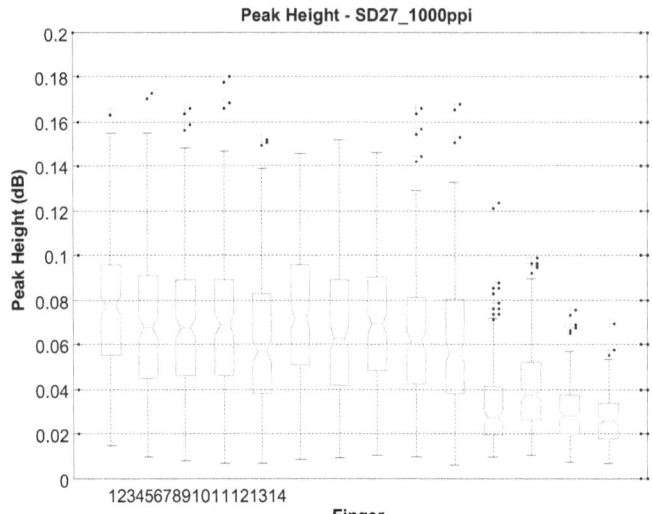

Fig. C-1 Box plot showing distributions of peak height ($dy$) by finger for SD27 dataset sampled at 1000 ppi. Fingers 1 – 10 are rolled inked prints on FBI 10-print cards; 11 and 12 are plain impressions of thumbs; and 13 and 14 are slap-four prints of left and right hands.

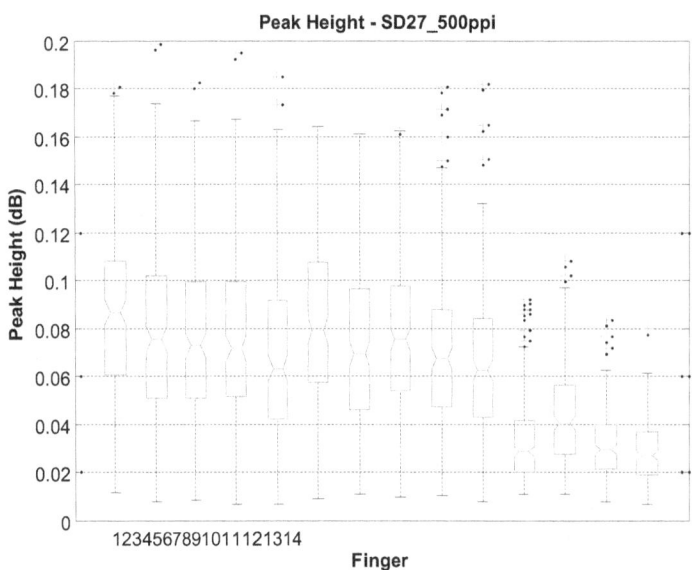

Fig. C-2 Box plot showing distributions of peak height ($dy$) by finger for SD27 dataset down-sampled (with anti-aliasing) to 500 ppi. Fingers 1 – 10 are rolled inked prints on FBI 10-print cards; 11 and 12 are plain impressions of thumbs; and 13 and 14 are slap-four prints of left and right hands.

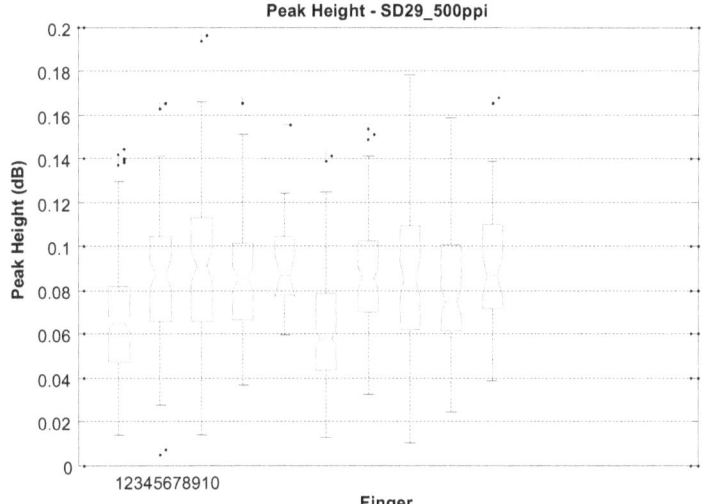

Fig. C-3 Box plot showing distributions of peak height ($dy$) by finger for SD29 dataset sampled at 500 ppi. Fingers 1 – 10 are plain inked prints.

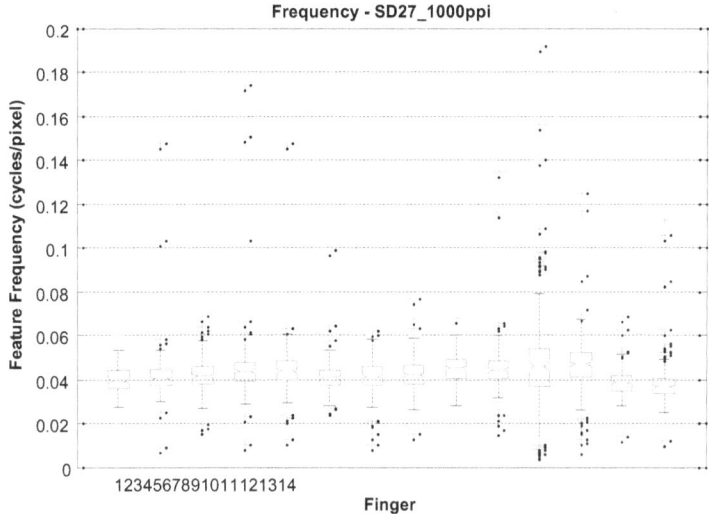

Fig. C-4 Box plot showing distributions of peak frequency location of the valley-peak feature by finger for SD27 dataset sampled at 1000 ppi. Fingers 1 – 10 are rolled inked prints on FBI 10-print cards; 11 and 12 are plain impressions of thumbs; and 13 and 14 are slap-four prints of left and right hands.

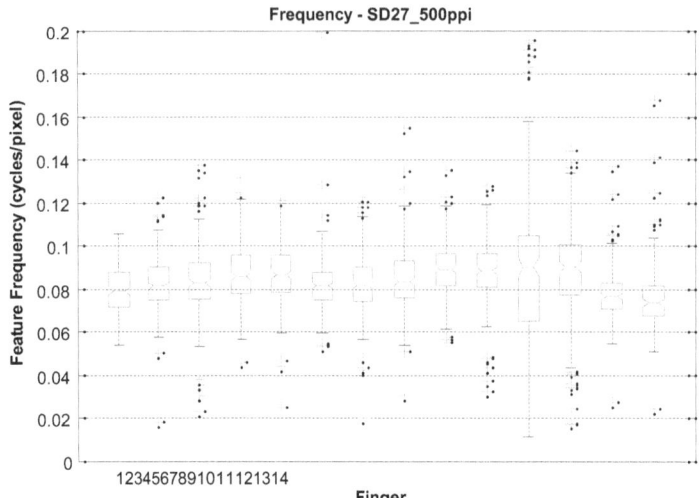

Fig. C-5 Box plot showing distributions of frequency location of the valley-peak feature by finger for SD27 dataset down-sampled (with anti-aliasing) to 500 ppi. Fingers 1 – 10 are rolled inked prints on FBI 10-print cards; 11 and 12 are plain impressions of thumbs; and 13 and 14 are slap-four prints of left and right hands.

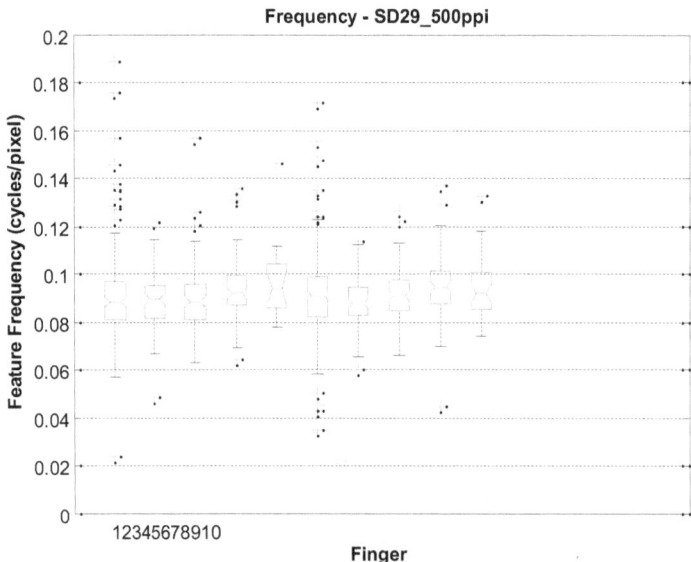

Fig. C-6 Box plot showing distributions of frequency location of the valley-peak feature by finger for SD29 dataset sampled at 500 ppi. Fingers 1 – 10 are plain inked prints.

www.ingramcontent.com/pod-product-compliance
Lightning Source LLC
Chambersburg PA
CBHW081746170526
45167CB00009B/3946